BORN FAE
The Inherited Gift

by Lorraine de Kleuver

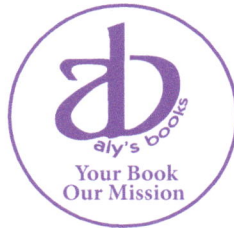

aly's books
Your Book
Our Mission

Dedicated to Kerri Brown, my cousin who
was always there when I asked the difficult
questions, of which there were many.

'Special Thank you' to Jenni Quinn
– for her design work in this book.

Born Fae The Inherited Gift
Copyright © Lorraine de Kleuver

First Edition 2025
Published by Aly's Books

www.alysbooks.com
Your Book | Our Mission

ISBN: 978-0-6459030-7-2

This is a global story...
This is a human story...
This is a supernatural story...

This will have you asking,
'Do we really know everything
about ourselves?'

Acknowledgements

A heartfelt thank you goes to the
following people who contributed
their experiences, in order to help those
who are new to connecting with
spirits and let them know
they are not alone.

Kerri Brown

Eileen Dal Bosco

George Dal Bosco

Kerry Dean

Alice Hudson

Jacobus de Kleuver

Shirley Mitchell

Contents

All shared stories are divided into sections, as per the type of connection(s) you identify with.

Introduction

I have to ask the question: *have we as humans discovered everything there is to know about human DNA?* Especially when it comes to things that can't be seen or measured within our body. Such as inheriting a supernatural gene, hidden within our DNA that gets passed down from generation to generation. The "gene" is something that enables spirits to pass through into our world. As far as I've experienced, it's one-way traffic; they visit us, not the other way round. Having said that, connections made between humans and spirits seem to occur to humans who possess something that a spirit can access.

Whatever the gene is, it's something that humans have had passed down to them from a time that is lost to us but still exists—dare I say—in a parallel world. It's not the first time the mention of a parallel universe has been referred to. In fact, it's practically old hat compared to the latest thinking in science, which talks about the existence of "multi-universes." *Google multi-universes for further information.*

However, I'm not a scientist; I am just an ordinary person who has experienced visitations from spirits and is trying to, number one, make sense of it all, and two, share the experiences I've had without sounding crazy.

In short, it would be impossible for me to even try because I can't, and I'm not the only one. There are many things that we humans can't explain. For example, we like to believe that we all have a soul, even though we can't see it, measure it, or study it, but we still strongly believe that we have a soul within us.

So, *why* is it so hard to accept that humans may inherit an ancient link that gets passed on through families, keeping humans connected to the past that remains ever-present? If we accept that familial traits are passed down, can we not also accept the unseen, unmeasurable, supernatural connection that lives within us as real, too?

However, for something to be real in this world, evidence is required. As much as I would like to supply that evidence in some academic paper, it's going to be super difficult—if at all possible. It would be easier if spirits cooperated and allowed us to take photos of them, but that's not going to happen. Although, I've been told many photos that reveal orb-like images of spirits are accepted (in the mainstream) as representing loved ones who have passed. The idea that humans have inherited a supernatural awareness of being Fay or Fae has been talked about for centuries; so much so that I can't understand why it's been dismissed as part of being human.

In an effort to answer this, I first have to look at my experience. For years, I was a person who thought people who attended mediums and fortune tellers were crazy. I didn't believe in any of that stuff—until I was up close and personal with several spirits in my bedroom one night.

Firstly, I wasn't on any psychedelic drugs, and my mental health was, and still is, in good order. However, try as I may to disbelieve what I saw, my brain overrode those thoughts. My brain held no doubt. In fact, my brain was pretty definite that all the spirits wandering around in my bedroom were real. Just as real as the cupboards, our beds, the wardrobe and the ensuite, excepting, of course, that the spirits were not made of flesh and bone – they were of the supernatural.

There was no going back after that night. After many visitations thereafter, I didn't question what I saw anymore—I believed them to be real spirits. I was not hallucinating, nor was I over tired. They were real, and what's more, they seemed to be preoccupied with what was in the room. As I watched them, they didn't appear to be in a hurry to see what was in the drawers, pulling out items and holding them up. It was like time didn't figure as a thing to worry about.

As I watched them study inside the drawers of the ensuite, they studied just about everything bar me. Thank goodness for that, as I would've screamed my head off. Alas, they weren't interested in me at all; I was incidental to what they were doing. When I think about it, it was more me studying them than them studying me.

After those visitations, I was a changed person, inasmuch as my attitude changed towards people who strongly believe in spirits, and that there is a supernatural world. What nailed it for me was finding out that my grandmother, aunties and cousins have inherited the gift of being Fae or Fay or both. When I sat down with them, I was surprised that they talked about it as if it were a normal, everyday thing. The aim of this book is to support the idea of gathering and sharing people's visitation stories and the connections made with spirits. *The Inherited Gift* focuses on what has been handed down and inherited through families for centuries.

It would be wonderful for libraries to dedicate a special space where people are encouraged to read and learn about such things as the inherited gift of being Fae, as well as the many and varied stories and practices that encompass the "Otherworldly". There is much more for us to learn and discover about the spirit world, and about connections made with spirits.

Having said that, there are hundreds of books about the Fae, predominantly being portrayed as Fairy Folk, which I find wonderful, yet I do not relate to them at all. The spirits I connected with were very much ordinary, hard-working people, but from a different time. What I mean by that is that their clothes and the type of employment reflected the era or century in which they lived and died.

What Does it Mean to Be Fae or Fay?

When I first looked into this, I found a lot of material about the fairy world of long ago, and those that are still supposedly with us. I wish a person of authority would come forth with factual information that is tangible and believable. It would be so much easier to quote from a source that is respected and widely recognised as an expert in the field of inherited fae or fay traits that are believed to come from the supernatural world. Alas, I haven't yet found that source.

All I can refer to is our family histories, and from what our DNA tells us about ourselves. Our DNA has been a real hero in finding out more about our ancestry, going as far back as when family records were collected and stored.

Unfortunately, when it comes to knowing our fae or fay traits within generations, it's more verbal stories that were kept and spoken about, but no data to support family members being Fae. I suppose they kept it quiet back in medieval times, in fear of being thought of as witches and getting burned at the stake. It would make sense to keep a low profile.

So, when it came to my own realisation that the spirits I saw—I now believe them to be real—was confirmed by extended family. What a joy and a relief it was, and that the gift of my being Fae was passed on from my paternal Scottish grandmother and (maybe) my maternal Irish ancestors who are recorded as living in County Clare a century ago.

My understanding of us being Fae (I believe) relates more to what we have inherited—which in itself has a bearing on how spirits connect to us—that enables some conduit or gene that connects the supernatural world to the human world.

For me, my gift enables me to connect by seeing spirits as they are: in their time and their space. Members of my extended family are enabled to hear, feel, have second sight, or be given heightened senses to feel a spirit presence is with them.

To be Fay is to foresee an imminent death that will occur. They do not know when, but they know it will happen relatively soon after the

fay prediction was given. My understanding is that the transfer of the prediction happens only when the Fay person has brushed or touched someone. Sometimes they know the person, and sometimes they don't.

I know of two people who have inherited the gift of being Fay; one has shared her story, and the other ended up opting out after discussing her experience on the phone only days before.

My understanding from listening to my cousin's experience of not being able to do anything, and having feelings of helplessness; I admire her trying to manage this burden as best she could.

I feel for both women, and even more so for the person experiencing being Fay for the first time and not knowing what to do. It would be frightening. I would hope that their mother or grandmother, from whom they inherited the gift, would still be around for them.

There are many books and websites that can reveal a lot more about being Fae or Fay. For me, it was explained that I was going through the "Awakening". I'm accepting of that explanation; I like to keep things simple and natural.

What is the "Otherworldly"?

Many kids' stories of yesteryear tell of magical happenings the fairies and pixies got up to. In effect, those stories are "otherworldly", where a child's imagination would be free to explore and dream of being in that "other" world.

I, myself, loved all my fairy storybooks. They took me to another place and time, just as watching astronauts on TV walking on the Moon did. As a teenager, I remember being emersed in the whole 'Apollo 11' journey of humans travelling in space. It certainly had me enjoying the subject of science at school a whole lot more. Coupled along with the American sitcom of the '60s called 'I Dream of Jeannie' had me totally ready for ghosts, spirits and space ships.

It should be of no surprise really, Jeannies have been around for centuries; associated with Islamic mythology, dating back before Christ. They were believed to be supernatural beings of the Otherworldly. For more info on historical accounts of Jeannies' I encourage you to search on Google. The history of Jeannies is truly amazing.

Now, being much older, I've had experiences that are definitely otherworldly, having me either waking up to figures wandering about or being woken up by the figures themselves, flicking my ear to wake me up. Those figures were spirits that appeared like ordinary people and were very real to me. So much so that at one time I screamed out loud, believing that two men were thieves in the night, robbing us.

After I pulled myself together and looked carefully at each spirit, I could see that they were from another space and time. This was recognised by the clothes they wore, which dated them, and the jobs they were still working on, such as two fishermen hand-rolling in large fishing nets manually. Aside from all that, none of them acknowledged me. It was as if I wasn't there. Nor did they acknowledge each other. The scene was indeed otherworldly.

In fact, when trying to work out what was happening in front of me, I was reminded of those pop-up gift cards that you get on a birthday. When you open it up, all the little cutouts pop up, appearing as if they're all together as one piece. However, when turning the card side-on, you can see the space between them, separating each cutout from the other. For me, it helped in grappling with the question of why they weren't knocking into each other and being oblivious to each other's presence. Plus, they seemed to be in a different time and space from each other.

Thinking more on that, if we accept that spirits are confined to their own space and time (i.e. the time they lived and died in), how can they then pass into *our* space and time? I can only defer to the supernatural connection that we have.

Why Do "Signs" Differ from Connections with Spirits?

It's a fair question. Well, so far, we have established that the inherited gift gets passed down through families, but not every member of a family has the gift of being Fae or Fay.

When I've referred to "connections" made with spirits, it seems that it's always on their terms. They choose to connect with us through a conduit—via our fae or fay trait— that they use when entering our world. Hence, we don't make appointments with them; they appear when they want to.

When talking about signs, I believe that the supernatural needs to be involved. Instead of sending a "message in a bottle", it's more like sending a message via the Supernatural world. The actual difference (as I see it) is that the messaging goes both ways. I mean, messaging via the supernatural space, is usually directed to a loved one who has passed.

For example, the loved one senses the heartfelt feelings and sends a sign in return. I've envisaged the process as follows.

Human   Having strong emotions and thoughts. Perhaps feeling down, maybe needing help, and really missing a certain loved one.

Spirit   Loved one's spirit responds to the emotions and thoughts that are being sent to them.

            The spirit selects something that they know their loved one will recognise as a sign sent specifically to them.

            Remembering that the supernatural world is closely aligned with nature, I imagine that a sign would be something from the natural world. For example, a butterfly from the garden, or a little bird appearing all of a sudden and hovering around you.

            Don't forget, the timely sight of a falling star or a breeze that plays with your hair. The sign is sent to be timely and close enough for their loved one not to miss it.

Human   Recognises the sign being from their loved one and immediately feels their presence with them (supernaturally).

Another wonderful thing I've noticed when people talk about receiving signs from loved ones is that this two-way messaging is available to everyone.

Signs can also be evidenced by other sources, such as poltergeists and other unknown entities of the otherworldly.

With poltergeists, it's more about seeing the results of their misbehaviour than seeing them.

Is it true? Well, as much as it all sounds a bit out there, the supernatural is out there.

If you refer to page 45, I have separated signs into two parts: messaging and visual evidence.

The Inherited Gift vs Predictions from Mediums

When considering the above, I can only fall back on my belief that all humans are born Fae. At the risk of repeating myself again, it's an inherited gift—centuries old. Which says to me that there is no versus anything, only differences in how spirits connect with us.

Mediums have the ability to **receive** messages from a spirit, as well as giving **voice** to spirits.

They are the most favoured 'go to' source for people that need answers to concerns. Whether they be past, present or future. Where I live, there is a Medium that has a three month waiting list, all via appointment so I've been told.

Mediums also help those of us who have the inherited gift of being Fae as well. Remember, having fae or fay traits doesn't mean that we all can foresee things. I can only connect with spirits via two connections; I can see them and feel the touch of a loved one.

I was in my early 20s, when it was a fun thing to do. The first prediction was from a woman who lived in Spotswood. She said that I would marry a man who sails the sea. A decade later, that was fulfilled, so I can tick that off.

The second prediction was from an elderly patient who was walking by me while I was writing some notes. She stopped and said, 'You're going to be a writer,' and some other stuff, as she kept walking with her wheely-frame. At that time, writing was not my thing. So, I took it with a grain of salt and forgot about it. However, a decade or so later, after writing and self-publishing many books, I think I can safely say that she was right. Another tick.

Mediums have long earned their place in their connections with the Spirit World. Evidence of their presence through the ages has them forever keeping us connected to the Spirit World, the Otherworldly.

The Gift

"Know that you are not alone in this."

The following stories and experiences
have been shared with you, so that
you may identify the gift you have.

It is the contributors' wish,
and mine, that you appreciate
and embrace your inherited
fae and fay traits that
connect you to the spirits
of the Otherworldly.

The Gift

When extended family and friends made it known to me that I was blessed with an inherited gift of connecting to the supernatural, that didn't mean that I could call upon the Spirit World as I pleased. It's rather more about spirits connecting with us, and that our gift makes that possible.

The gift of being Fae or Fay is like inheriting a specific key or gene that accepts specific messages from spirits, allowing them to enter our space and time. Through your gift, your eyes are opened to the supernatural world. In saying that, you still have your bed and house on Earth and your family with you.

In my experience with visitations from spirits, it appeared to me they were curious; they were interested in what we've got that they haven't. I won't lie to you; I didn't know what the heck was going on in my first encounter. After the initial shock of seeing them and screaming my head off, I gradually learned to cease the screaming and quietly watch them. I never watched them for long. In fact, I don't remember falling asleep. However, I did become a bit over-vigilant, sleeping with one eye open and holding the sheet half over my head as if to show them that I was asleep. Stupid right? I'm embarrassed to admit it to you.

So, don't worry, we all react differently when we're not sure of what to do.

This book is designed to be a resource for those who have just discovered their inherited gift, at a time when it's all new to them. Some who read this might not have a clue about what's happening to them. Your gift can remain dormant for years, or, for some, it can make itself known when they're quite a bit older. My experience of visitations didn't occur until I was in my 60s. Usually, your family or extended family know about the inherited gift and can explain who else in the family is a bit Fae or Fay. As I mentioned earlier, I have several cousins who are Fae, all of whom had it passed down from our grandmothers respectively.

If you're a newbie to all this, I can say you are not alone. This book aims to provide helpful information and advice from people who have inherited the Gift. Their shared stories hope to offer great comfort and help you to embrace this special gift you've inherited.

Manifestation of your Gift

When trying to sort out in my mind what I've heard and have been told about, I was to try and make sense of the gift that myself and many others like yourself have inherited.

The following categories, as listed on the next page, are some of the many ways people have connected with Spirits. Note that Fae is the common Fae trait; with the other Fay spelt differently, as shown on the following chart.

Once you've identified the connection that you've experienced, you can flip through to that section.

Note: It is possible to inherit both being Fae and Fay, which you will read about. If it seems a bit overwhelming the following stories will help you through it.

Known Types of Connections

Audio connection
You can *hear* spirits moving about.   Fae trait

Empathic connection
You can *feel* or sense a presence.   Fae trait

Heightened Senses
Have a strong *awareness of a presence.*   Fae trait

Second Sight connection/premonition
Foresee an imminent accident/bad news.   Fae trait

Visual connection
You can *see* spirits, as if they are real people.   Fae trait

Vocal connection
You can *speak* to spirits.   Fae trait

Foresee a death connection
Foresee the imminent death of a person   Fay trait

Connection via a medium
Medium works as a go-between spirit and human.   Fae Trait

Signs connection
Via use of **messaging** or via use of visual evidence   Fae trait

Connection with a *hostile spirit*   Fae trait

As we come ever closer to accepting and understanding that we have always been together, having the Supernatural close at hand.

With our loved ones and our kinfolk, we will rejoice hand-in-hand, when we finally meet up again with our never-ending spirit families.

– LDK '25

Shared Stories & Experiences

After identifying the Connections made with spirits, we encourage you to select the section that you identify with, and read through the shared stories from people who have had the same or similar connections.

Know that their stories are to remind you that you are not alone. You will be believed. You're not going crazy.

There's a lot more to us that we don't know about, as there's a lot more to know about our world. Keeping an open mind helps us to accept not only the power of the supernatural world, but also the power of the supernatural within us.

–LDK '25

Connection 1 – Can hear spirits

Name: Kerri Brown

Connections with: Her mother

Gift: Experienced both hearing and touch

Handed down from: Maternal female line; mother, grandmother

Heritage: Geordie (UK Newcastle)

Note: Kerri is also Fay. Please refer to Section 7 to read her experience of being Fay.

EXPERIENCE

Kerri's mother, Jean, was known to have the second sight. A standout story that the whole family remembers came about one morning when Jean's husband, Jim, was getting ready for work.

He worked at the coal mines in Wonthaggi. While Jim was readying himself for work, Jean had asked him not to go. When Jim asked her why, Jean said she just had a feeling and that he should stay home. As Jim continued his preparations to leave, Jean, who is usually a calm woman, gradually became more agitated. As Jim started to leave, Jean picked up a brick and threw it through one of the windows in the kitchen to force him to turn back.

It was no surprise to her that a section of the mine where Jim would've been working had collapsed that day. Thanks to Jean's second sight, Jim escaped what may have had him incur life-threatening injuries.

From then on, it was accepted in the family that Jean's feelings, however they presented, were to be acted upon. Jean had a stoic belief that our loved ones walked with us after they died and had influence over our decisions.

Later in Jean's life, she was to have a near-death experience from complications arising after surgery. While receiving a blood transfusion, she left her body and headed to the light. She spoke of how the light filled her whole vision, and she was in a peaceful, calm place. Jean said it was a beautiful feeling, but it was not her time, and she was pulled back into her body. Kerri explains that her mother said that she never feared death after that experience.

Kerri herself speaks of her own empathic experiences as a teenager, being aware of events or happenings that were to come. One memory she recalls is when her grandfather was in the hospital. She didn't know what he was in hospital for, and as far as the rest of the family was concerned, he was expected to make a recovery. Kerri asked a friend to stay with her at the hospital because she knew that he was going to pass away that very night. His passing was a complete surprise to the family, as he wasn't particularly old.

Years later, when Kerri lost her own parents, she felt her father just sitting on the end of her bed. Kerri remembers always feeling peaceful when he was there. She remembers that, although they didn't communicate with each other, she would have that peaceful feeling wash over her. Then she would drift off to sleep.

Her mother, on the other hand, was more tactile. Kerri explains that she would often hear her mother's distinctive footsteps around the house and follow her journey from room to room, checking on us. At times when she was crying in bed, mourning her loss, she would feel her mother brushing her hand over her hair to calm her. A soothing habit she had used over the years to assure Kerri that everything was alright.

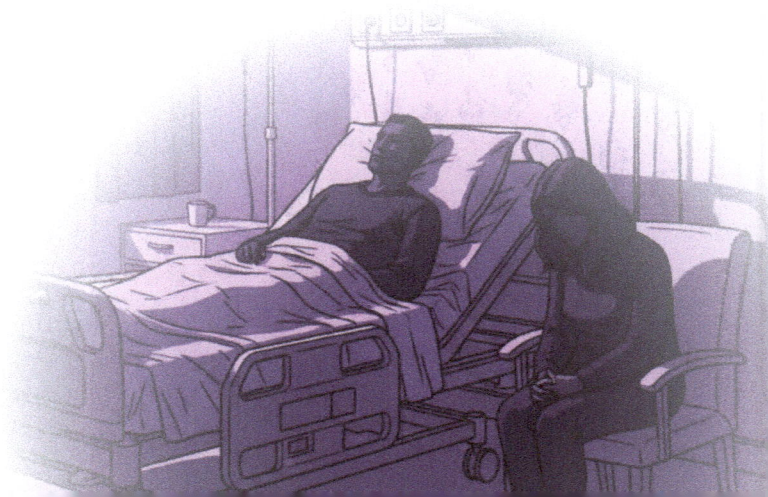

Name: Hannah (Eileen's grandmother)

Connection with: Her son

Gift: Hearing spirits

Handed down from: Maternal/Paternal line via grandmothers

Heritage: Irish

EXPERIENCE

Post the death of Hannah's son to a brain tumour.

Eileen's grandmother used to say that she could often hear the opening and closing of the chest of drawers in her son's bedroom. Even though it wasn't actually happening, she knew that it was him returning.

Note: Further to Hannah's story, she experienced her son's presence around her after his death.

Please refer to Connection 3, (page 29).

Name: Lorraine de Kleuver

Connection with: Mother

Gift: Hearing a presence

Handed down from: (Maternal) mother/grandmother

Heritage: Scottish/Irish (County Clare)

EXPERIENCE:

More recently, on May 20, 2025, I was woken up in the middle of the night to the sound of loud knocking on the front door. At first, I sat bolt upright, not sure what to do. The three knocks sounded heavy and loud, so I worried that a tall, heavy bad guy was waiting on the other side of the door. The thought of calling out to ask who was there didn't enter my mind, nor did opening the door. I just stood in the dark thinking of what to do. Realising that I didn't hear the sound of the wire door or the steel gates being opened, my thoughts dwelt on the possibility of spirits trying to send me a sign of sorts.

Convinced that it was spirits, I went back to bed, wondering what the sign would be. As I started to fall back to sleep, I thought someone I know might have died. Satisfied that that was probably the case, I fell asleep. When morning came, my mind thought about my stepsister and wondered how she was. She was quite a bit older than me, and I didn't have a lot to do with her after she left home, got married and had a family. She was busy with her own life and I'm happy to report that she is alive and well.

Later that morning, I opened Google and typed in a question about the phenomenon of knocking on doors in the middle of the night, delivered by Spirits.

Guess what? Apparently, it's common. There were many other explanations, but I was happy to settle on it being a common phenomenon.

> Dreams release us from this world
> into the supernatural world, where
> we can feel free.
>
> – LDK '25

Connection 2 – Can feel spirits

Name: Lorraine de Kleuver

Connection with: Mother

Gift: Feeling her presence

Handed down from: (Maternal) mother/grandmother

Heritage: Scottish/Irish (County Clare)

EXPERIENCE

I was in my mid-20s when I experienced the sensation of being hugged by a spirit. The hug felt strong and reassuring; it was a feeling that I welcomed. I can remember thinking to myself that it was my long-passed mother.

At the time, I was enduring a really painful earache. It was Friday, and I just managed to drive myself home from work. My earache had got worse during the day. I got a couple of Panadol during the day, but later that afternoon, I asked if I could leave early as the pain worsened.

By the time I got home and walked up the stairs to my flat, I felt I could no longer stand and had to sit down. From then on, my earache would throb greatly at the slightest movement of my head. The pain was excruciating. It was then that I received the strong feeling of being hugged. It was about 8 at night. Beforehand, I put the telly on to distract me from my earache. I had taken another couple of Panadol before sitting down. As I tried to position myself in the chair, keeping my head still, I must've fallen asleep.

Forty-seven hours later, I woke up. My earache was totally gone, and I felt great. What was a mystery to me was the memory of being hugged. I don't remember going to bed at all or turning the telly off. I remember nothing but sitting in the chair, holding my head and being in severe pain.

I still remember the hug even now, and how it felt. There was an intervention of some kind, because I was incapable of moving at all — perhaps mum and her spirit friends placed me in bed and decided to stick around for a bit.

Connection 3 – Sense a presence

Name: Aunt Lily (Eileen Dal Bosco's aunt)

Connection with: An awareness of spirits being around her.

Gift: Sense a presence

Handed down from: Mother/grandmother

Heritage: Irish (Thurles, County Tipperary)

EXPERIENCES

As told by Aunt Lily's niece, Eileen.

Sensing the presence of a Spirit was a common occurrence for Lily. In fact, Eileen recalls that her aunt was not at all afraid of them.

Thoughts from the author

STOP right here, please! Let's take a breather.

It's so easy for us to pass over this because humans are used to having that feeling of something we can't see, hear, smell or touch. Yet, we can sense a presence. In other words, we can feel it. Those feelings can range from having the hairs stand up on the back of your neck to getting a hug from a deceased loved one.

Now, if I told you that sensing a presence is just some unexplainable thing that we humans have gotten used to—and who cares anyway, right?

Up until now, I've been referring to the supernatural as inhabiting the otherworldly; but that's only one space the supernatural inhabits. When thinking about our fright, flight and fight hormones alerting us to danger – I have to ask, what actually alerts those hormones if there's seemingly no evidence to be reactionary to anything.

There's a lot of talk about the supernatural being within us – and it's more than just talk. There's plenty of evidence to back it up. I suggest you visit your wonderful local library, or the Information giant, that is Google.

To add a bit of humour to this; when finding yourself in a dark, scary lane, remember you have an inner supernatural watch-dog that is barking at your hormones to RUN!

For Aunt Lily, sensing the presence of a friendly spirit will always be welcome to her. I believe the supernatural sends both friendly and warning messages to our bodies in ways that we can respond accordingly.

Aunt Lily was also subjected to signs.

To read more, turn to Section 9, "Connection with Signs".

Name: Hannah (Eileen's grandmother)

Connection with: Her son

Gift: Sense a presence

Handed down from: Maternal/paternal line via grandmothers

Heritage: Irish

EXPERIENCE

Eileen tells Hanna's story, and what she said after her son died of a brain tumour. Hanna spoke about how strongly she sensed her son's presence around her when he passed away.

He would've been my Uncle Tom. He died in his 20s. Although I never met him, I've heard stories that, when he was alive, he would often open and shut all the drawers of his chest of drawers. I think it was when he got confused about things due to his brain tumour.

Nanny used to say that she could often hear the drawers opening and closing. Even though it wasn't actually happening, she knew that it was him returning.

Note: The above has been mentioned in Section 1, "Hearing of Spirits".

In all the countries around the world,
there are people that speak of a destiny,
a place that has no beginning and no end.
It's a supernatural world where only spirits dwell.

LDK '25

Connection 4 – Second sight/ foresee/premonition

Name: Alice Hudson

Connections with: Her mother

Gift: Experienced both premonitions and outcomes

Handed down from: Maternal female line, mother/grandmother

Heritage: Geordie (UK Newcastle)

EXPERIENCE

My elder brother Jim was a reserved young man, deeply committed to his academic pursuits. With an innate sense of responsibility, he naturally became the protective guardian of his younger sisters, always ready to confront those who dared to antagonise them. Within the walls of our church, Jim stressed the significance of decorum, imparting lessons on respect in a sacred space. He emphasised the importance of honouring our surroundings, reminding us that any misstep in behaviour would not go unnoticed and would undoubtedly be addressed after the service concluded. Ultimately, guided by these values, Jim chose to dedicate his life to the church, pursuing a vocation as an Anglican priest.

There was one notable occasion when Jim invited me to accompany him to Inverloch. He needed a companion for a meeting related to his nomination with the Australian Labour Party. As an unmarried man, I often found myself at his side for various dinners and gatherings, an arrangement that felt quite natural. Initially, I agreed to join him without hesitation.

Yet, as the day approached, an unsettling sense of apprehension washed over me, a feeling I couldn't easily shake. I finally confided in Jim that I would not be attending and urged him to reconsider his plans as well. Unfortunately, he brushed aside my concerns, his determination unwavering, as he set off on his trip.

Tragically, fate intervened during his return journey. Jim was involved in a harrowing accident, crashing into the rear of a tractor on the twisting Inverloch Road while making his way to Wonthaggi. The aftermath was swift; he was rushed to Wonthaggi Hospital by ambulance, where he received treatment for multiple injuries, including a fractured chest bone. The days that followed were spent in the hospital's sterile environment, under observation for the various traumas he'd sustained.

Another instance where I failed to heed my intuition occurred on the eve of my marriage. An overwhelming feeling of dread clawed at me, whispering that perhaps I should forgo the ceremony altogether. After bravely discussing my concerns with my mother, she surprisingly agreed with my trepidation. Yet, my father held a different view, believing that the wedding must continue as planned, and thus, the ceremony proceeded.

Upon our return from our honeymoon in Lakes Entrance, we faced a series of unforeseen complications. My spouse found himself behind the wheel of a car that was not his own, thanks to some mischievous friends who had tampered with his vehicle. In a remarkable show of generosity, Jim lent us his car for the trip. Despite my growing reservations about our journey, I expressed my concerns multiple times, advocating for a delay. Yet, my inner apprehensions were silenced, and we set off on our way.

Tragically, as we navigated the Strzelecki Highway near Morwell, fate dealt us a cruel hand. We collided with an unmarked council vehicle engaged in road repairs, resulting in a sudden and jarring impact due to the lack of warning signs. In a stroke of luck, I had instinctively slid down in my seat during the collision, as seat belts were not yet mandatory. The aftermath was catastrophic for the vehicle, reducing it to a mangled wreck, barely above the height of the front seat's headrest. Miraculously, I escaped with my life, was unconscious and woke up in Mowell hospital under observation. Still, the incident left an indelible mark on my psyche and my forehead, filling me with an overwhelming sense of regret, as to this day, I have been haunted by guilt for not having stopped the ill-fated trip.

Another instance that remains etched in my memory occurred during a drive from Wonthaggi to Dandenong with my spouse and a young baby in tow. Again, the whispers of dread echoed in my mind, urging me to postpone the trip. Common sense wrestled with my apprehensions, but ultimately, my concerns were dismissed.

As we descended toward Pakenham, our car skidded dangerously on black ice. Though we escaped unharmed, the fear of what could have been loomed large over me. Later, on the way home, as we stopped at a set of traffic lights in Dandenong, disaster struck once more. A sudden, horrifying crash erupted from behind us—another vehicle had barrelled into us, failing to stop. The driver sped away, leaving us in a daze as we inspected the damage. We pursued him until we found his car parked outside a vehicle repair business in Somerville. It turned out the driver was the owner of the shop; he had been at a funeral and had indulged in a few too many drinks. He offered to fix our car at no cost if we agreed not to involve the police, a deal we accepted, largely because we weren't certain about our insurance coverage. Fortunately, no one sustained serious injuries, but the memory of that day lingered, a reminder of the weight of my unheeded instincts.

In the wake of these life-altering experiences, I have grown increasingly attuned to those oppressive feelings that arise unexpectedly, recognising their importance. I now understand the necessity of acknowledging these instincts and adjusting my plans or direction accordingly, always prioritising intuition over the noise of everyday life.

Name: Shirley Mitchell

Connections with: Her mother

Gift: Premonition

Handed down from: Maternal female line; mother

Heritage: Irish

EXPERIENCE

In this story, Shirley speaks of a premonition that struck her while on holiday.

It was when my husband Phil and I were driving up to the Blue Mountains to have some time out and restore ourselves after a busy period. We decided to take a trip to the Blue Mountains in New South Wales and visit the Three Sisters rock formations.

For some reason, I felt a bit anxious just before my body jolted. I felt I couldn't go any further. These feelings of anxiety occurred when we began our climb up the Blue Mountains. I had a premonition. I called out to Phil that there was something wrong. "I don't know what it is," I said. "But there's something definitely wrong."

I stopped the climb and phoned to make sure the kids were okay. Then, because I wasn't feeling great myself, we decided to forget about the climb and go to a restaurant for dinner on the cliff. As we began to walk back, I got a phone call from an old family friend, Teresa.

Teresa was looking after Mum and had taken her up to a little country town called Berry, not far from Wollongong, for the day as an outing. While walking around the shops, Mum had a cardiac arrest. An ambulance was called in time for mum to be resuscitated successfully. After hearing what happened to her and coping with all the emotion that my premonition created, I was done. When we arrived back at our bed and breakfast, I burst into tears and told Phil, 'I can't do a thing.' Despite it happening years ago, just talking about it now still brings tears to my eyes. Alas, after that crisis, Mum lived to a ripe old age of ninety-three years—she was amazing!

Connection 5 – Can see spirits

Name: Lorraine de Kleuver

Connection with: Spirits

Gift: Seeing spirits

Handed down from: Maternal/paternal line via grandmothers

Heritage: Scottish/Irish

EXPERIENCE:

When they appeared to me at our home in Sale, I screamed out loud when I saw two male figures that appeared to be looking into our drawers. As soon as I screamed, they began to make their way to the front door. My husband didn't react to my scream until I hit his feet a couple of times. My husband has both his hearing aids out at night, so initially didn't hear me. Before he turned his bedside light on, I saw enough to make out that they were fishermen trying to roll up their fishing nets as fast as they could. They wore old-style caps and clothing, early 1900s style, if I were to take a guess.

Of course, they were gone once the lights were on, and I was back to explaining what I saw again. That was only the beginning. The next visitation was having a crowd in the bedroom.

It wasn't enough to see a 6-foot-plus Mauri warrior walk out from my wardrobe. He was one big fella, wearing a handmade, painted skirt that looked like strips of bamboo sewn together. He had long dark hair that was wavy, but flowing in ripples. I couldn't help but notice that he politely moved the wardrobe sliding door back until it was closed, before walking towards the bedroom door. Even then, he had to bend to prevent knocking his head

While watching him, he passed an elderly man, who looked like he was hanging on a cross, wearing a red cape over his shoulders and a bright white gown over his body. He had very grey-white shoulder-length hair.

While I was looking at him, he was frowning back at me. When I tell you this, I can say that at no time did I think he was Jesus, more like a protester of some sort wanting to be Jesus. While that was going on, an older lady was bending over my husband, trying to look at his face. Firstly, I thought it was his mother because she was a small lady and had a round tummy. Then, when I saw her short hair, I realised it wasn't his mother because she always had her hair in a bun. While all this was happening, not once did I feel afraid. I just watched them for a bit and went back to sleep.

As much as they have their effect on me, having my sleep interrupted, they didn't have the same effect as a young boy did. It was weird how it happened. One minute I'm asleep in bed, then all of a sudden, I sat up and called out, '"You're going the wrong way!' As I watched this young teenage boy walking in the darkened space, I remained sitting up, wondering where he was. Carefully, he guided his way through what seemed like a long tunnel. And, God only knows how I knew that he was going the wrong way!

He didn't hear me, so I called out again to him that he was going the wrong way. He was wearing faded green shorts and a faded red t-shirt. His hair was styled like that in the '50s, short hairstyle with a part to one side. I only saw the back of him. I watched him for a bit, then I went back to sleep. I remember all of them—him especially, because I couldn't help him.

I also noticed that, despite the spirits walking in different directions as they pleased, not once were they aware of each other, nor did they bump

into each other. It was as if they were in their own protective bubble, which kept them from communicating. They were in their own time, their own era, in a state of contemplation, or looking for something.

The only one that wasn't contemplative was the protester fellow; he was frowning at me the whole time. So, I had enough of him and went back to sleep.

Even though I had no fear of them and had stopped trying to wake my husband up, I felt that I had taken just about enough of them. My sleep was being interrupted, making it difficult to concentrate during the day.

Even the most enjoyable experiences of being woken by something flicking my ear to make me open my eyes had become tiresome. I would wake to see countless gemstones before me in the most beautiful colours, shapes and sizes, spiralling upwards towards the ceiling. I watched in awe as they glittered in the darkness.

I remember looking up in wonder, moving my hands in and around the gem stones. I wasn't frightened at all; in fact, it didn't even feel strange. I saw no hand throwing them upwards; they just rose into the darkness. When it was over, I went back to sleep with a smile on my face, feeling happy from the experience. That has only happened on two separate occasions.

It wasn't long before I found a solution to ending my interrupted sleep by spirits. When shopping at Aldi's, wandering along their centre aisle, I saw a range of nightlights for sale.

That night, I plugged my one-and-only nightlight into a power point in the middle of the hallway. Since then, I have had no more visitations or interrupted sleep.

Sometimes I think about removing the nightlight, but I remember wishing that they would go away. Yet, I still wake up every so often to see if any spirits came to visit. Now, that is crazy.

Name: George Dal Bosco

Connection with: His pet dog named Boz

Gift: Seeing his pet dog

Handed down from: Mother/Grandmother

Heritage: Italian

EXPERIENCE

The story goes, as told by his wife, Eileen.

I remember George telling me when his dog Boz, a black Staffy, died years ago. Two weeks after his passing, George was sitting in the lounge when he saw Boz walking past him. He was just heading from the lounge into the kitchen, just like he always would.

I myself have sensed my dog Pip (a Cavalier) lots of times. I think anyone that has had a pet in their life are the same. The feeling is fleeting, just long enough to make you turn around as if they were there.

Name: Eileen Dal Bosco

Connection with: Male spirit

Gift: Seeing spirits

Handed down from: Mother/grandmother

Heritage: Irish

EXPERIENCE

Eileen remembers a time when she was a child in England.

I was in my bunk bed—the bottom bunk—and my sister was in the bed above. I looked over to the doorway where the landing light was on. I saw a figure of a man just standing there. He was a hazy white colour. I also remember getting very scared and wanting to yell out to my mum and dad, but I was scared that he might come over to me. So, I put my head under the blankets, and, in my mind, I said, 'Please go away, please go away, please go away.' I fell asleep under the blankets, and that was the end of it.

I also remember a time involving my son, Shaun. He often went to look out the big lounge room windows and talk to someone at night. I used to think someone might be out there at night, trying to get friendly with him, planning to grab him. Either that, or he was talking to spirits. Anyway, I was listening to late-night radio 3AW one night, where a priest who used to be on the show was featured. People would ring up and speak to him about anything they wanted. I remember this night, a lady rang up and spoke about a ghost in her house, saying she was scared. The priest told her to just keep repeating, 'Please go away.' He told her whoever it is will leave her alone. So, I thought I'd try this myself, in case it's a spirit speaking to my son. So, I did, and it never happened again.

Connection 6 – Communication to spirits

Note from the author.

I'd would like to say that I've experienced chatting with spirits, but I haven't, and neither has any family members or friends that I know of. However, I have heard of spirits possessing a human, in order to use their voice to communicate through.

Oddly enough, I think it may have happened to me.

I was attending to my usual habit in the mornings of pushing blinds back from the windows and doors. My husband was with me, when I said to him, "I'll just push these blinds back," (at least that's what I thought I said). What came out of my mouth was gobble-de-goop.

When hearing the words spilling out, it sounded like it was a rea different language, delivered with certainty and without hesitation. My husband clearly heard me and gave me a confused look. I laughed it off. ¯hen when I tried to have another go, the same voice with the same message was repeated as if I knew the language fluently.

We both stopped and looked at each other. Thankfully, it ended soon after and I was back to my normal self. I laughed it off saying "don't worry, I haven't got dementia yet." Sometime later, when I was a one with my thoughts, I knew that I had spoken in another language. At a guess it sounded African, not that I would I really know. That type of experience with spirits has never happened again since.

Although this page is titled 'Communication to spirits,' I would've loved to have had a talk to a spirit, and be able to tell you all about it. I must say though, that this spirit was doing her level best to try, albeit I wouldn't have understood a word of it. "Oh Well"!

Connection 7 – Foresee a death. Fay

Name: Kerri Brown

Connection with: Foreseeing a Death

Gift: Having both Fae and Fay Traits

Handed down from: Maternal line heritage Scottish/UK (Jordie)

EXPERIENCE

Kerri is my cousin. Aside from her grandmother, mother and one of her older sisters being gifted with fae traits, Kerri was the only one in her family gifted with both fae and fay traits.

Born into a large and loving family, Kerri was able to share her experiences with her mother. Her being Fay revealed itself when she was a young teenager. She knew when her grandfather was going to die, when the rest of the family expected that he would get better. I was a teenager when I went to visit him with family; he was up and dressed, sitting at the end of his bed, acknowledging the extended family who were around him.

Her next predicted foresight was the hardest to bear: that of her own husband's passing. This cut to her very core.

For people who are Fay, having received a premonition of an imminent death, must be like torture. One has no choice but to accept what has been foretold, knowing that there's nothing they can do about it and that it will be fulfilled. Worse still, one is not given how or when the premonition will take place; they only know it will happen.

Kerri's husband was a roofing plumber. He fell from the roof of a commercial building. Rendered unconscious at the scene, paramedics had placed him on life support before transferring him to the hospital. Sadly, he passed away six days later. Two weeks prior to his passing, Kerri woke from a dream with a feeling of dread. She woke up her husband,

and, with an emotion-filled voice, she said to him, 'What will I do without you?' When he saw the look in her eyes, he tried to comfort her by saying, 'Don't be silly, I'm not going anywhere.' That night, and the nights that followed, she carried the burden of complete helplessness alone, in silence.

When a premonition is made, it will be fulfilled.

Connection 8 – Received Prediction. Given by a medium

Name: Lorraine de Kleuver

Connection with: Medium in Spotswood, Melbourne

Outcome: Medium's prediction came true after ten years.

1st Prediction: The medium predicted I would marry a man that sails the sea, and I would sail with him.

This prediction was totally unexpected and true. I didn't believe the medium's prediction. After so many years, I had forgotten all about it. Despite suffering seasickness, I sailed with him on many occasions.

Connection with: Medium in Werribee.

2nd Prediction: The medium predicted that I would be a writer.

Extra notes: This prediction was absolutely unexpected and true. Again, I didn't believe the medium's prediction, especially when I had no interest in being a writer.

She was an elderly lady who stopped by my work desk. Placing a finger firmly down on the desk without looking at me, she revealed her prediction of what was to come.

Also, the prediction took about ten years to be fulfilled. I have self-published twelve books, ten of which were children's books.

What started my writing journey? I believe it all started when I began looking into my family history. After listening to people's stories and putting them all into a book, I felt that I had achieved something good and worthwhile. I've found that at the core of why I love writing so much is the process and the people I meet along the way.

Connection 9 – Signs

Types of Signs

MESSAGING

This form of communication involves the otherworldly ways of producing two-way messaging, meaning that human thoughts of despair, and uncertainty, feelings of hopelessness are being picked up by a loved one from the supernatural world.

The loved one in spirit picks up those thoughts and sends a sign to lift their spirits, showing them that a loved one is thinking of them.

This type of messaging is available to everyone, whether you are Fae, Fay or not.

VISUAL EVIDENCE

Spirits engage with us in a myriad of fascinating ways. It's not uncommon for spirits to leave evidence of their visits. Whether it's by leaving an indentation or shape behind, for example, on a chair that nobody usually sits on, or cushions being thrown on the floor all the time. At first, you may not recognise that the sign is left deliberately, until the same sign is repeated again and again, for months on end. For a friend of mine, the same sign was repeated consistently for years.

For the most part, the items that get moved around may be ones that hold memories for you. Signs are as engaging as they are bewildering, such as poltergeists being mischievous and leaving a mess behind. On the whole, whenever I hear about someone receiving a sign, it's primarily believed to be from a loved one.

Name: Lorraine de Kleuver

Type of Sign: Visual Evidence

HAVE YOU HEARD OF POLTERGEISTS?

Well, some parapsychologists view poltergeists as a type of ghost or supernatural entity that is responsible for psychological and physical disturbance. Others believe that such activity originates from "unknown energy" associated with a living person or location. Sceptics, on the other hand, prefer mundane explanations such as attention-seeking, pranks and trickery.

Source: TheConversation.com, German Folklore. Noisy Spirit, Wikipedia

MY EXPERIENCE

I was in my early 20s when I worked as an enrolled nurse at the Queen Victoria Hospital. I was on night shift in Ward 5 East. This was the kids' ward. I was partnering with an RN named Deb Brown.

The Queen Vic was situated right in the centre of Melbourne on Swanston Street. I had worked a six-month stint in Ward 5 West—the babies' ward—before working in the kids' ward. So, I knew both wards and the routines pretty well. However, on one particular night, the usual routine turned out to be very, very different.

Unlike other hospitals, Queen Vic was a bit noisier outside, due to the trams going up and down Swanston Street and the occasional ringing sound when someone wanted to get off at their stop. As well as the trams, there were the patrons pouring out of nightclubs and being loud about it.

It was after midnight. Deb and I had completed our rounds, checking the kids and filling in documentation. The ward was quiet. So, after we got our cup of tea each, we both settled back at the nurses' station. The nurses' station had clear views of all the kids in the ward; we couldn't have missed anything. Although the ward had low lights, the station was close enough to see or hear anything amiss. Apart from the sounds from the lift well doors opening and closing in the corridor, all was well.

That was until I heard the sounds of things falling on the floor. Quickly, I grabbed a torch, wondering if one of the kids had thrown something out from their bed. After doing all the bed checks, everything was fine. When returning to the nurses' station, I shone the torch where there were no beds. To my surpise all the kids' books were strewn over the floor. I knew I had tidied around there and couldn't understand what had happened.

After I put them all back on the bookshelf and returned to the station, I told Deb what had happened.

"I can't understand," I said to her. "I know I tidied all around there."
"It's probably poltergeists," she said.
"What's that?" I asked.
"It's excess energy, coming from the kids" she replied matter-of-factly.
"Never heard of it," I said, as I went to sit down.

After getting another cuppa, I sat down in what was a peaceful ward again. But it happened again. Reaching for the torch, I shone the light towards the bookshelf. Again, all the books were scattered as if the whole bookshelf had been turned upside down.

I quickly went back to Deb, saying, "It's happened again." She too got hold of a torch and went around the ward. While I put all the books back, she checked all the kids, the beds and everything else. Any dark corner that looked like something could be hidden there had a light shining towards it.

After double-checking everything, we both turned the front of the bookshelf to face the wall, leaving a torch shining on the bookshelf. There was no more funny business with the bookshelf that night. I was amazed that all the kids slept through it. Perhaps that's because it was just books; if they heard the sound of a timber bookshelf hitting the floor, things might've been different.

Name: Kerri Brown

Type of Sign: Visual evidence

EXPERIENCE

The following was emailed to me by Kerri, as outlined below, about her older sister's passing.

"Did I tell you about Susan's funeral day?" she asked.

Sometime before her passing, but knowing her death was imminent, she asked her sister Alice if she thought heaven was real (this from the most spiritual person I knew). On being told the affirmative, she then asked, "Do you think I'll get in?"

Alice replied, "If you don't get in, then there's no hope for the rest of us." Susan said she would send a sign. They decided on a lightning bolt.

Susan was a great organiser, and she had instructed her hubby to take all the boys to the pub between 5 and 6 p.m. We three sisters were sitting around the poolside at the motel just winding down.

At about two minutes to 6 p.m., the sky darkened, and at exactly two minutes later, a single straight lightning bolt appeared. We all just looked at each other and said, "Well, she made it!" The sky cleared straight away after that, and we were happy knowing she was in a better place with our parents.

It still gives me goosebumps.

Name: Kerry Dean

Type of sign: Visual evidence

EXPERIENCE

The following story is from my girlfriend, whom I have known since my secondary school years.

I've never known her to talk about signs, spirits or anything relating to those subjects. It was many years later when visiting her and her family that the subject of spirits arose. She began talking about a particular chair in her lounge room. At that time, I wasn't a believer in ghosts or spirits.

Anyway, she mentioned that this particular chair had been sat on every night for years. Before going to bed, Kerry always gives things a little tidy up, and as part of her routine, she pats down the cushion on that chair.

Whether her visitors believed her or not, she swears that she knows some spirit or other sits in that chair every night. It didn't bother her, but she wondered if it was a loved one who had passed away keeping an eye on the house when everyone was asleep. Kerry's story is one of a sign that was evidenced over many years. The sign was seeing the imprint of someone having sat in the chair overnight.

A couple of years later, after the passing of her husband, she and her son, Tom, moved to another place, but still close to where her mother and extended family live. She hasn't yet said anything about having that sign occurring again.

Name: Aunt Lily (Eileen Dal Bosco's aunt)

Type of Sign: Visual evidence

EXPERIENCE

Eileen recalls her aunt telling her that when a picture fell from its hook and onto the floor, it indicated that someone she knew had died, or that something bad was going to happen, like receiving bad news.

In addition to Eileen's recollections of experiencing signs from spirits is as follows:

Name: Eileen Dal Bosco

Type of Sign: Visual Evidence

EXPERIENCE

When walking down a street while holidaying in Ireland, the sight of my mother's Irish maiden name, Condon, was displayed on a window of a funeral parlour. At the time, I thought it was unusual, as it wasn't a common name. Further, on that day, I sighted a red robin in a tree and went up to look at it, thinking it was very strange that it didn't try to fly away, no matter how close I went to it. So, I took a photo of it and the funeral parlour.

Later, when my husband and I were waiting at the airport to board our next flight to Italy, my brother (back home in Australia) rang me to say that my mum was very unwell and had been transferred from her room in the nursing home to palliative care. He told me it would be best to come home if I wanted to see her before she passed away. I have since heard that Red Robins are messengers from God when someone is dying or has passed away.

Following on from Eileen's story, where she touched on her seeing a red robin and the symbolism they hold with the otherworldly, I've added a few more: blackbirds, crows, owls, ravens, robins and wrens.

Connection 10 – Being Threatened by a hostile Spirit

HOW TO SEND A MESSAGE TO UNWELCOME SPIRITS

Most connections with spirits, from stories I hear about, are without any dramas, but an elderly lady, well into her 80s, tells of a story full of torment.

Her tormentor is her late husband, who regularly wished terrible things upon her, keeping her awake. Nobody deserves that sort of trauma, but she is a woman who has a powerful faith in God. Her faith helps her to cope.

Unfortunately, as I understand it, her family were of no help at all; they didn't want her to talk about their father in unflattering ways. I suspect she has some friends that she can vent to, without worrying her children, who are adults.

I'm not a part of her inner circle, but I have since found that having a nightlight on outside the bedroom has helped me when I wanted to put a stop to any more visitations (I was getting cranky from interrupted sleep).

Also, I learned of other options such as, smudging one's home with white sage essential oil sticks. As I mentioned earlier, this means going from room to room repeatedly communicating to spirits (respectfully) they are not welcome in your home.

Caveat: Any connections made with spirits of the otherworldly are at their pleasure. If and when they wish to reveal themselves, it is at their discretion. I'm afraid you cannot simply summon a spirit to visit you when you want. If you feel the need to talk to them, do so with great care and respect, so as not to make the mistake of offending them.

OLD FEARS AND SUPERSTITIONS

When putting this book together, I came across what I believe to be old fears and perhaps superstitions from a time long past that can still have a hold on people today. Evidence of this was presented to me when I asked people if they were interested in contributing a story for this book. Initially, they were interested.

However, for some, when it came to having their name sitting a ongside their story, problems arose. Some felt they didn't want to risk upsetting the spirits. Hearing something like that brought it home for me; despite the modern world we live in, fears of the otherworldly are still quite real. Faced with this, I endeavoured to find a possible reason for those responses.

Protecting one's image

Some strongly believe in spirits of the otherworldly, but don't want to be seen or heard as believing in them.

Reputational challenges

When people change their perception of you from a solid, grounded person to a person who is talking about spirits, it can be a challenge for them to take you seriously.

The paradox of being Christian and being Fae

I believe in God, but not the church. I believe in the power of nature. She surrounds me, sustains and protects me. I believe in the supernatural otherworld, I believe in the Supernatural that is within me, and the healing power it provides. I believe all humans are born Fae.

The real paradox here, is that to be born Fae is to acknowledge that Christ had something to do with it. He is the Creator of everything: including the supernatural otherworldly.

> "I realise that writing this book is a risk for me and my reputation. There are certain people in the world, including some in the science community, who might call my work pseudoscience."
>
> Dr. Joe Dispenza
> Doctor, Scientist and modern-day mystic.
> Author of 'Becoming Supernatural' How Common People Are Doing the Uncommon, Wall Street Journal bestseller, New York Times best-selling author

THE SUPERNATURAL: OUT FROM THE SHADOWS

When I reflect on when I wasn't confident in talking about my experiences, doing a bit of research showed that I was not alone. Aside from extended family explaining a few things to me, doing a bit of reading was really helpful.

I have to say that Google opened up a whole new world in information gathering. I wanted to read and find information from the world I live in—the modern world. Once I found the approach that I wanted in my writing, I found the right questions to ask Google. From then on, I was amazed and happy that the supernatural wor d is occupying the brains of many academics, involved in careers such as the following:

Anthropology: The study of human societies and cultures and their development.

Metaphysics (Ontology): Ontology is concerned with the study of existence and reality itself.

The meaning of Ontology: The term comes from the Greek word for "being". So, it's the study of being, or the nature of being. Most people describe the "ontological Turn" as the realisation that there can be more than one ontology, more than one theory to explain the nature of things.

EXCERPTS FROM TWO UNIVERSITY PAPERS

It was satisfying to read *Taking the Supernatural Seriously (Academically Speaking)*, by Sofie M. Hansen (2017), a student journalist and anthropology student at the University of Copenhagen. The supernatural world was getting serious attention.

Sofie talks about anthropologist Morten Axel Pedersen, who often runs into situations that, at first glance, appear paradoxical.

'What do you do when you're doing field work in Mongolia, and you meet a shaman who also happens to be the head of a company that has been given a World Bank loan? Or what about when you find yourself at the University of Moscow talking to someone getting their PhD in quantum physics, and they ask if you'd like to talk about what to do when you find yourself in the mountains, surrounded by spirits?'

I've read Sofie's article a few times now, and feel it reinforces the strongly held beliefs in the existence of a supernatural world that continue today.

In the abstract to his paper, *Naturalism, Science and the Supernatural*, Steve Clarke (2009) writes:

'There is overwhelming agreement amongst naturalists that a naturalistic ontology should not allow for the possibility of supernatural entities. I argue, against this prevailing consensus, that naturalists have no proper basis to oppose the existence of supernatural entities. Naturalism is characterized, following Leiter and Rea, as a position which involves a primary commitment to scientific methodology and it is argued that any naturalistic ontological commitments must be compatible with this primary commitment. It is further argued that properly applied scientific method has warranted the acceptance of the existence of supernatural entities in the past and that it is plausible to think that it will do so again in the future. So naturalists should allow for the possibility of supernatural entities.'

link.springer.com
https://.springer.com/article/10.1007/s11841-009-0099-2cs
Article: Naturalism, Science and the Supernatural I SPHIA
Published: 24 APRIL 2009

WHAT ABOUT THE PHENOMENON THAT IS DÉJÀ VU?

One can talk about this subject for ages. When reading up on it, it seemed every man and his dog had an opinion on it, including scientists and so forth. Theories of the phenomenon being of the supernatural and of the paranormal got my attention. However, that little voice in my head had me second-guessing.

My eyes fell on two simple lines of text, as set out below:

'Déjà vu is caused by a brain misfire that creates a false sense of familiarity. It's usually harmless but can be linked to fatigue, stress or seizures.'

When I had my first and only experience of déjà vu, I was in grade one at East Oakleigh State School. My mother had passed away. My siblings and I had moved from one extended family to another in Wonthaggi until our father found a foster home for us. I remember being told not to be a sook; people don't like sooks. I was to call the lady at the foster home mum". I never did; I called her by the name as she was introduced to me: Mrs Jacobs.

I didn't want to be there, and I didn't like the foster home. After a few weeks, I remember heading back towards the classroom. The atmosphere around me started to become hazy. When I sat down at my desk, I had this overwhelming feeling to lie my head down on the desk for a moment. My teacher at the time, Mr Hammer, called out to me to sit up a couple of times, but I couldn't. Then my whole body started to shake convulsively. I couldn't stop. He picked me up and raced me to the head teacher's office. When I woke up, I was in bed back at the foster home.

After that, I refused to eat, which made Mrs Jacobs very angry with me, which in turn made me more determined to die – yep, if mum was in heaven, that's where I wanted to be.

Not long after that, I experienced a déjà vu event. Even now, at sixty-eight yearsold, I remember it so well, maybe because I had an incredible uplifting feeling with it. It was summer, and Mr Jacobs had put the sprinkler on the front lawn. We were running out towards the front door as an adult couple was entering. I ran past them and stood on the veranda, watching the other kids laughing as they ran in and out from

the sprinkler. I remember being happy. It was a beautiful morning. While I was standing watching, I experienced a strong feeling that I had been on that veranda before, in that house, watching the scene before me. I stood still and fully took in what was happening to me. It was only a very short time, yet it felt longer. As soon as the experience passed, I told one of the teenage girls what had happened. She smiled and said, 'Sounds like déjà vu.'

I really don't know how it happens or what brings it on. It might well be due to fatigue and stress. Despite science discovering a lot more about stress and the many ways it presents itself. Nevertheless, I am not convinced that stress in itself is an acceptable explanation for the occurrence of déjà vu experiences. Dare I say that de'ja' vu is more relatable to the supernatural otherworldy events.

That simple moment of joy may have contributed to me allowing myself to have fun. We stayed there for four years, and in that time, I began to make friends. One friend I remember well was a girl named Diane Barns, who lived very close to the school. She was fun to be with, and so was her family.

SUPERNATURAL HEALING

The more I delve into Spirits and the Supernatural, the more I find it all so incredible. What I once believed to be old knowledge and stories from the past are thriving and very much relevant in the modern world. Take, for instance, self-healing.

In saying this, I'm not referring to taking a Panadol and she'll be right, mate. This is far, far bigger than I ever could imagine. I'm talking about serious healing of the human body, inside and out—supernaturally!

There are a tremendous number of scientific books and papers on this subject. Unbeknownst to me, the concept of healing one's own body has been around for some time with great success.

Hearing about it and reading about it is one thing, but when my husband told me that he has been healing himself supernaturally for years, I was really surprised. A little bit of background about my husband: he was a theatre nurse at St Vincent's Hospital (Melbourne) and Warragul Hospital.

CO'S STORY OF SUPERNATURAL HEALING

This story is as told by Co himself:

It was some time after my first wife passed away. I was still feeling depressed and out of sorts, not sleeping well, and constantly trying to work out why it happened.

I decided to try to help myself get some "proper" sleep by relaxing and telling myself that I was getting better. Every day thereafter, I continued to repeat to myself the words, 'Every day in every way I will get better and better!' Surprisingly, within a couple of nights, I began to feel the results, sleeping better and longer without the constant flood of memories.

It became a routine for me, and I kept it going. I also tried to focus on some problems I was having at the time. When I developed indigestion problems, I kept telling myself during my nightly meditation that my inner self would repair and fix this problem. Within a short time, it had eased considerably and is now no longer a problem.

Author's note on self-healing:

Since sharing his story with me, and my reading a book about being in the supernatural, I am learning that our brain does a lot more for us than just being a busy organ; it's our go-to place for self-help.

The supernatural has always been there for us; what's different about then and now is that science has results. They have the statistics and the successes to back it up. Our inner self is the supernatural domain within us; our brain is the control centre that receives the instructions sent via the repeated practices of meditation and acts on them. It is a lot more complex than that, but that's my understanding of the process.

It is a never-ending journey; the more you read, the more you learn, and the more you learn, the more you want to know.

When searching for more information, I tend to seek simple explanations that are easier to understand and are more enjoyable reading material. I began to ask myself why I was getting so fixated on the supernatural when there wasn't really anything tangible about it. For starters, what can you measure? It's not a gas, liquid, or solid. You can't breathe it. At least our astronauts have displayed that by having to wear special space suits to protect them and stay connected to oxygen. So why do I, and the rest of the human race, bother about wanting to learn about it at all? The following page provides a simple and satisfying explanation—well, at least for me, and I hope for you, too.

> ## "Supernatural thinking is actually an important part of being a complete human being."
> Dr Clay Routledge,
> in partnership with the Templeton Foundation
>
> ### THE TEMPLETON FOUNDATION
> Sir John Templeton lived in an era of unparalleled scientific and technological progress. The accelerating pace of scientific discovery led Sir John to wonder whether the methods of science might be harnessed to make similar progress in understanding the deepest and most perplexing questions facing humankind. Today, the Foundation that bears his name aspires to fulfill his vision — to create a world where people are curious about the wonders of the universe, free to pursue lives of meaning and purpose, and motivated by great and selfless love.
>
> Referenced: www.templetonfoundation.org

WHY OUR MINDS CRAVE THE SUPERNATURAL

Ever since "The Enlightenment", society has become more rational and less religious; in particular, "supernatural thinking" has declined.

However, according to existential psychologist Dr Clay Routledge, supernatural thinking is an important part of being a complete human being.

When we ponder great existential questions—like our purpose or place in the universe—our minds naturally drift toward the supernatural. The supernatural provides hope that there is something about our existence, something about the human spirit, that transcends the material.

Enlightenment, a European intellectual movement of the 17th and 18th centuries in which ideas concerning God, reason, nature, and humanity were synthesized into a worldview that gained wide assent in the West and that instigated revolutionary developments in art, the philosophical and politics. Central to Enlightenment thought, were the use and celebration of reason the power by which humans understand the universe and improve their own condition. The goals of rational humanity were considered to be knowledge, freedom, and happiness.

(Source: Wikipedia)

GOALS FOR CHANGE

Achieve a sense of normalcy: when discussing inherited Fae or Fay traits. I believe this would go some way in helping people to talk about it, especially for those who find it scary and confronting to connect with spirits for the very first time.

Encourage talking about it: Get the family around and bring out the old family photos. Share in the experiences and the stories. Continue the handing down of stories from the past.

Be Respectful: How would you go about bringing up the subject to friends, knowing that you may risk being laughed at or not believed?

I have been that person, the one who didn't take what a friend had to say seriously. That day, I had lost a friend. She never spoke of it again. Just know that when a friend confides in you about their experience in connecting with spirits, please don't do what I did.

Remember, this is a serious matter that should not be taken lightly. For a person to have the courage to share this with you means that they trust you. Please encourage them to talk about it, perhaps take them to a quiet place so the conversation can be held in private.

Having said that, I personally think it's better to reveal it to family first. It is a family thing.

If I could go back in time, I would be a better person and ask my friend to talk about it, with a mindset to hear them out. I would try to show I believed them, to help them feel comfortable about being open about it.

My experiences, and learning from my extended family that I have inherited the gift of being Fae, helped me to understand the strange things Nan used to do when I stayed with her as a kid. I have her name as my middle name: Agnes. When I think of her now, I wonder if she is witnessing all this (in the comfort of the other world) and happy that I have received what she, her mother and all the mothers going back in time have passed on.

As for my mother's side, I know little about the Scottish/Irish heritage, other than names and dates. For me, it equates to having half of oneself missing. This makes it difficult to know their stories. Remarkably though, through all the ravages that human history holds, this one unseen, unmeasured, and unstudied family trait (or supernatural gene), our inherited gift, has survived *everything*.

I wonder, could the following short question-and-answer quiz show that inheriting the gift is really this simple?

Q: When humans inherit attributes from their parents, where does it all happen?
A: Within the mother's womb, where the parents' gene makeup combines.

Q: When does being Fae begin?
A: As soon as you are born and have taken your first breath.

Q: At what age should you expect your "awakening" to occur? As a child? As a teenager? As an adult?
A: Any one of the above.

My Awakening became active at five. Christmas was days away. It was a warm night. I awoke in the middle of the night. The atmosphere in the bedroom became very cold. I sat up in bed, not wanting to go to the toilet or wanting for anything. I just woke up with a sense that something bad had happened. My mother had died that night. I believe that her spirit came into the bedroom (I guess, to say goodbye), before parting to the Otherworldly.

In my early 20s, I experienced feelings of her presence, especially when I was unwell.

Then, in my early 60s, spirit visitations became full-on, taking over the bedroom.

In my late 60s, suffering lack of sleep caused by spirit visitations, I noticed they all would vanish the instant my husband turned his bedside light on. Tired of being woken up at night, he bought a nightlight for the hallway. It's been three years now, and I have been enjoying restful sleep ever since.

In Closing

When I began this book, it was to help others when faced with spirits for the first time. Although there are many books on this subject, I wanted to write about my own experiences and those of extended family and friends.

During the writing process, I couldn't help but feel that the spirits were guiding me with this book. When I felt disheartened and doubtful in my ability to complete it, they would present an idea that would make itself known to me, a word that would kick-start me back into writing, or an idea that would change the direction of my thoughts. It was nothing I could feel or sense; I knew I was getting help from them, but it was more than that.

The spirits instilled a strength of mind within me, protecting me from getting overwhelmed by it all. I'm now comfortable in believing that they're real, and that where they dwell in the supernatural, other world, is real. Just as the inherited gift that I was born with is real, so it is for you.

Postscript: The Human Soul

** Not familiar with meditation, your local library can help you.

Last but not least, I can share with you something truly stunning.

When I began this resource book, I was asking the questions to try and find out more about my Inherited gift of being born Fae. When I got to the stage that I had shared as much as I could, along comes something special, something that science has revealed: the location of the human soul. The following excerpt reveals what was originally unknown to me.

> 'In the fourth energy center, we are moving from being selfish to being selfless. This center is associated with emotions of love and caring, nurturing compassion, gratitude, thankfulness, appreciation, kindness, inspiration, selflessness, wholeness, and trust.
>
> 'It is where our divinity originates; it is 'the seat of the soul.' When the fourth energy center is in balance, we care about others and we want to work in cooperation for the greatest good of the community. We feel a genuine love for life. We feel whole and we are satisfied with who we are.'

Excerpt from the book Becoming Supernatural. How Common People Are Doing the Uncommon. Author: Dr Joe Dispenza. Getting Better Acquainted with the Energy Centers.

Don't Stay Isolated – Create a Fae Group

I can understand that it's easier said than done to create a group, despite there being plenty of fae groups on Facebook and the like. After a decade of flicking through posts, giving "likes" and "loves", I've decided to set out and create a group that I can actually sit down and have a coffee with.

The main aim is to form friendships where people can share their experiences and talk openly about their inherited gift. Sure, we have our families to share with, but not everybody does. This connection would provide newbies with a go-to place where they can hear stories from others and feel comfortable sharing their own.

If you wish to talk further, I can be contacted via email at dekleuvers@gmail.com.

'In spiritual traditions, the hummingbird is often seen as a bridge between the physical and spiritual realms. Many indigenous cultures believe that hummingbirds carry messages from ancestors or divine beings, offering guidance and wisdom.

Beyond rituals, the hummingbird teaches valuable lessons in mindfulness and presence. Its ability to move swifty yet remain focused on a single flower serves as a metaphor for living in the moment. In a fast-paced world, the hummingbird reminds us to slow down, appreciate the beauty around us, and find joy in the simple things.'

Selected excerpts from:
'VacWarBirds.net.' Animal Symbolism.
Hummingbird Symbolism: A Journey Through Myth,
Meaning, and Modernity.

Caution: Any connections made with spirits to the Otherworldly, is at their pleasure. If and when they wish to reveal themselves it is when it suits them. I'm afraid you cannot just summon up a spirit to visit you when you want. If you feel the need to hear from them, do so with great care and thoughtful consideration as not to make the mistake of offending them.

Perhaps it's best that you seek out a Medium if you feel it's urgent.

www.ingramcontent.com/pod-product-compliance
Lightning Source LLC
Chambersburg PA
CBHW040908210326
41597CB00029B/5020